27 SOUTH FULTON

ATLANTA FULTON PUBLIC LIBRARY

SCIENCE AT WORK

USING MATERIALS

Eric Laithwaite

Franklin Watts
London   New York   Toronto   Sydney

© 1987 Franklin Watts

First published in 1987 by
Franklin Watts
12a Golden Square
London W1

First published in the USA by
Franklin Watts Inc
387 Park Avenue South
New York, N.Y. 10016

First published in Australia by
Franklin Watts Australia
14 Mars Road
Lane Cove
NSW 2066

UK ISBN: 0 86313 481 5

US ISBN: 0-531-10264-5

Library of Congress Catalog
Card No: 86-50352

Printed in Belgium

Designed by Ben White

Illustrated by Hayward Art
 Group

Photographs:
Associated British Ports 5
Beneteau 17*l*
British Aerospace 7, 28
British Alcan Aluminium Ltd
 13*l*
British Telecom 17*r*
Colorific 26*r*
Daily Telegraph Colour
 Library 4
Len Dance Associates 10
Chris Fairclough 8,11,18,19*l*
Glasshouse Crop Research
 Institute 19*r*
ICI Fibres 6
Kos 27
Magnet Southern 14*r*
NEI Parsons 13*r*
Pilkington 9
Renault 26*l*
Rolls-Royce 16
Science Photo Library
 7,28*t*,29*r*
Shell 22,23,24
Tony Stone 15,20,24*b*
Vauxhall Opal 14*l*
Wedgwood 21

Contents

Natural materials	4
Manufactured materials	6
Three states	8
Metals	10
Uses of metals	12
Wood and paper	14
Glass and ceramics	16
Water	18
Gases	20
Materials that grow	22
Fossil materials	24
Chemicals	26
Modern materials	28
Glossary	30
Index	32

Natural materials

A material is a substance that we use, such as the paper in this book. Natural materials are materials that we find in the world.

The most common materials on Earth are air and water. We tend to think of air as being other than a material because it cannot be seen or touched as can water. But air *is* a material just the same. It can be weighed, and has many uses.

We can see the effects of this invisible material best when the wind blows. It moves anything light which is not fixed to the ground. It bends the stems of plants and the branches of trees, just as if a jet of water were being aimed at them. Being less dense than water, its effects are not so dramatic as those of a hosepipe, but they are easily recognizable as having a similar cause.

△ Air enables things to fly. A hot-air balloon has a burner that heats the air inside. Because it is less dense or lighter, hot air floats in cool air just as wood floats in water.

▽ Even though it is invisible air greatly affects movement. When light objects such as feathers or pieces of paper fall, they float down gently because they have to move the air aside. A parachute works in this way.

Materials other than air and water which we find on Earth include soil and sand, rock, wood and leaves, plant fibers such as cotton, ores, coal, gas, crude oil and animal products such as wool and feathers.

We have found an amazing range of uses for natural materials. Most of the objects in a house, as well as the house itself, have been made from materials that have grown or have been dug out of the ground. It is a strange fact that apart from salt, everything we eat has, at one time, itself been alive or part of a living thing.

We have also learned how to make new materials from natural ones, using the science of chemistry. When we make large quantities of such materials, we are doing chemical engineering.

▽ Natural materials are not always found exactly where they are needed. Not all of the sand and gravel needed to build a city comes from the ground on which the city will stand. This ship is unloading building materials through a pipe. Water is a most useful material for transporting heavy loads from place to place by sea, lake, river or canal.

Manufactured materials

Clay is a natural material found in the ground. When shaped and baked in an oven it becomes brick or pottery. This material, called a ceramic, is an artificial material made by people. Most metals are not used in their natural state; the majority of them have to be extracted from ores. Paper and cardboard, are the result of processing the natural material, wood. Natural fibers such as cotton and wool are spun and woven to make cloth. Glass is made from sand and other natural materials.

The contents of a modern room may make more use of processed materials than of materials in their natural state. Materials such as plastics are made from chemicals that come from natural sources, especially oil, and are often called synthetics. They may be better in some ways than natural materials. Clothes made of synthetic fibers do not need ironing; plastic cups do not break easily.

△ Nylon, polyester and other synthetic fibers are cheaper than natural fibers and often stronger and harder wearing.

▷ All these articles are made of artificial or processed materials, including fiberglass (for insulation), semiconductors (in the electronic circuit board), and plastics and synthetic fibers such as polystyrene and nylon. The metal cans contain engine oil and gasoline. Even the "wood" is actually plastic.

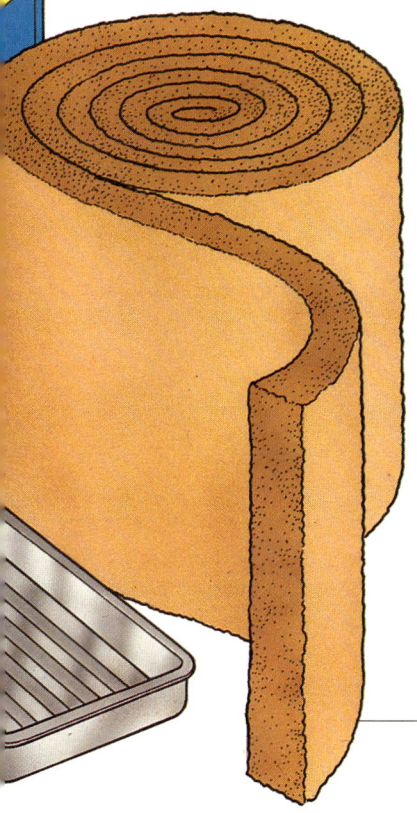

△ Engineers working on Giotto spaceprobe. The demands of space science have led to completely new types of material.

Just over a hundred and fifty years ago, our whole way of life was changed dramatically by the discovery of how to use magnetism and electricity. Special materials were needed and these were initially of two kinds: conductors and insulators. Those through which electricity can flow easily we call conductors. Most metals are good conductors. Materials that resist the flow of electricity are called insulators. They include plastic and ceramics.

Electricity was first used to make heat and light. Then came electric motors, which provide the power for movement of machinery in factories, in the home and for electric railways.

Electricity made possible such inventions as the telephone, radio and television, tape recorders and computers. Cars, aircraft and spacecraft all depend on electrical circuits. Developments in electrical knowledge led to great advances in surgery, food production and communications. The quality of modern life has been improved greatly by the use of magnetism and electricity. Yet new advances in these subjects are still being made every year.

Three states

△ A burning candle shows all three states of matter. The solid candle holds a pool of liquid which turns into gas as it burns.

Most materials can exist in three forms: solids, liquids and gases. These are called the states of matter. Solids tend to resist any attempt to change their shape. Liquids and gases, on the other hand, have to be kept in a container or they would flow away. Generally it is easy to change the state of a material from one form to another. When we make use of materials, we often do this.

For example, solid glass can be melted into a liquid and poured into a mold to make new shapes. So can most metals. Water can be frozen into a solid (ice) or boiled to make steam in order to provide cold or heat.

The state of a material depends on the movement of the tiny molecules of which it is made. In a solid, the molecules move to and fro in very small vibrations. In the liquid state groups of molecules can move around. In a gas, the molecules have complete freedom to move independently even though they may still collide with each other.

△ Ice is made up of crystals in which each water molecule is held in place as if in a framework made of wire.

△ Ordinary water contains water molecules, which are free to move around but tend to stay together in small groups.

△ In steam, the water molecules are spaced far apart and move around almost independently of each other.

We can change the state of a material by changing its temperature or the pressure to which it is subjected or both. Air and other gases can be made liquid by cooling them under high pressure. The "natural" state of a material is usually taken to be the one that it adopts at room temperature and atmospheric pressure. Chemically, a change of state may occur simply due to the mixing of two materials. Adding salt to ice makes the ice melt.

Many solids form as crystals in which the molecules line up in certain patterns. This organized molecular structure gives crystals their regular shape. Sugar, salt and ice are all common examples of crystalline material. Metals too, are made up of crystals whose structure makes them strong and hard.

▷ Large, very flat sheets of glass are made by pouring hot liquid glass on top of a bath of hot liquid tin. The glass stays floating on top of the tin as it cools. Here the glass is being washed after cooling. Glass made in this way has a very smooth surface.

Metals

Metals are among the strongest and toughest materials. We use them for making objects that have to last a long time and are subjected to heavy wear due to rubbing, beating, or the effects of weather or other hostile environments. Metals keep their strength in conditions of high pressure, high temperature and powerful forces, such as are found in engines and in buildings. They make the gear wheels for mechanical clocks and watches, and the wheels, chassis and bodywork of cars and trains.

Metal cooking pans conduct heat rapidly to the food they contain. Washing machines, refrigerators and similar household machines are generally encased in steel sheeting to give firm protection. Silver, gold and platinum have a beautiful metallic luster that makes them valuable in jewelry.

△ An important use of a very strong metal such as steel is for reinforcing concrete in buildings. Although it is very hard, concrete is not very strong when subjected to heavy loads. Steel rods which are set in place before the liquid concrete is poured, hold and pull the concrete together so that it resists strong forces.

◁ Welding, here done by a robot, joins metal parts by melting their surfaces together. The two parts become one solid metal object when cooled.

△ Engineers check over a big turbojet engine. Many metals combine strength with lightness. This combination is used best in the aerospace industry.

Some metals such as iron, copper, aluminum, lead and tin are pure metals extracted from ores. Others are mixtures of two or more pure metals, and these are called alloys. Brass is an alloy of copper and zinc. Bronze is a mixture of copper and tin. Alloys can be made only by first heating the metals until they become liquids and then mixing them together.

Aluminum is the most common metal. Over ten per cent of the Earth's crust is aluminum. It is also one of the lightest metals.

For well over a thousand years after we discovered how to extract metals from natural ores, only a few pure metals such as copper, tin, lead, zinc and iron and two alloys, brass and bronze, were in common use. It is only in the last century that their range of usefulness has been extended enormously by alloying common metals with rarer metals such as nickel, chromium and magnesium.

Uses of metals

Metals are used to make many objects simply because metal is stronger than any alternative material. Metals usually have more strength than wood, plastics and cement, for example.

However, if unprotected, many metals combine with the oxygen in water, causing them to corrode or rust away. Some rare metals, such as gold and platinum, are much more resistant to this corrosion. Because they are expensive, they are often used to put a thin, protective coating on more vulnerable but cheaper metals.

Most metals conduct electricity extremely well, and iron and steel can be magnetized. Steel and copper are essential to all electric motors and generators and copper and aluminum make up over 95 per cent of all electric circuits. Without electricity, many modern machines would not work.

△ Metals are used in electrical machines not only because they are good conductors, but also because they are convenient. Metals can be easily joined by soldering, are readily cut, and can be coated with insulating materials with which they do not react chemically.

▽ Hot liquid aluminum is poured into molds and allowed to set to form solid bars.

△ The amount of power that an electrical machine such as this generator can handle has been increased by over 10 times in the last 100 years. A part of this increase has been the result of new additives, notably silicon, in the iron cores of the electromagnets. Permanent magnets have also been improved by new alloys of iron containing cobalt, copper, nickel, aluminum and titanium.

Tungsten is a pure metal with a very high melting point which can be used for the filaments of light bulbs without melting. Cadmium is a relatively cheap metal that is used as a non-corroding plating material for other metals. Mercury is liquid down to temperatures well below the freezing point of water and so can be used in thermometers.

A small amount of carbon added to iron makes it into steel, one of the hardest, strongest and therefore most used metals. Similarly, a small amount of nickel and chromium added to steel makes stainless steel, which does not rust.

Some metals can be joined by using soft solder – a mixture of lead and tin with a relatively low melting point – which binds with a polished surface of copper and brass. A tougher joint is made with brass as a solder and can be used to join iron.

Wood and paper

Wood is softer than metal and therefore more easily shaped using cutting tools and abrasive tools like saws and sandpaper. Unlike metals, wood cannot be melted, and for this reason all wooden objects have to be cut from solid pieces of wood. On the other hand, wooden parts are more easily joined together, using various joints, a variety of glues, or simply screws or nails.

The surfaces of wooden objects can be protected and made attractive by polishing, staining, varnishing or painting by applying liquid material that either soaks into the wood or hardens on the surface.

Because it is a material that is cheap and easy to handle, yet strong, a lot of wood is used in building a house. It forms doors, window frames, roof supports and floors, and some houses are built entirely of wood.

▽ Wood is easily transported in the form of planks which can be tied into bundles. It can also be transported as large sheets.

◁ In developing new types of aircraft or road vehicles, it is very useful to test small models in an airstream within a wind tunnel. These models are made by gluing together layer upon layer of wood. The outer surfaces are then highly polished or coated with metal sheeting.

We use many products that are made by treating wood. One method of processing wood is to slice off thin sheets of wood from logs. These sheets are then glued together face-to-face to make laminated wood or plywood, which is very strong and yet flexible.

Another wood product is made by pressing cheap wood chippings and glue between two sheets of high-quality wood, sometimes coated with plastic materials. It is often used for shelves and work-tops.

Paper and cardboard are made from wood that has been made into a pulp — almost a liquid — which is then spread out into thin sheets and dried. Paper and cardboard are among the most commonly used processed materials. The shelves of supermarkets are stacked with cardboard packages.

There are many kinds of paper. Some, used in cooking, are non-porous and will contain grease without absorbing it. Paper towels are porous but retain some strength when they become wet so that they can be used for cleaning.

△ Newsprint is the word used for the very cheap quality of paper used for newspapers. Discarded paper can then be re-pulped and used again.

Glass and ceramics

Glass is made mainly from silica, which is found in sand. Mixed with limestone and soda at a high temperature, it becomes a liquid. As it cools, it becomes a soft solid. In this state it can be molded into shape to make jugs and vases. Alternatively it may be poured as a liquid into molds to make bottles or on to flat surfaces to form sheets of glass for windows, mirrors and greenhouses. Glass is also used to make lenses for cameras, telescopes, microscopes and eyeglasses.

Some glass can be made tough enough to be bullet-proof, and port holes of very thick glass enable observers in submersibles to see the sea bed at great depths where the glass has to resist enormous water pressure.

▽ The turbine in a jet engine may have to work at temperatures hot enough to melt metal blades of any kind. These ceramic blades can withstand great heat.

▷ Many objects in the home are made of glass or ceramics, and the building itself may be constructed of brick, a ceramic, with glass windows.

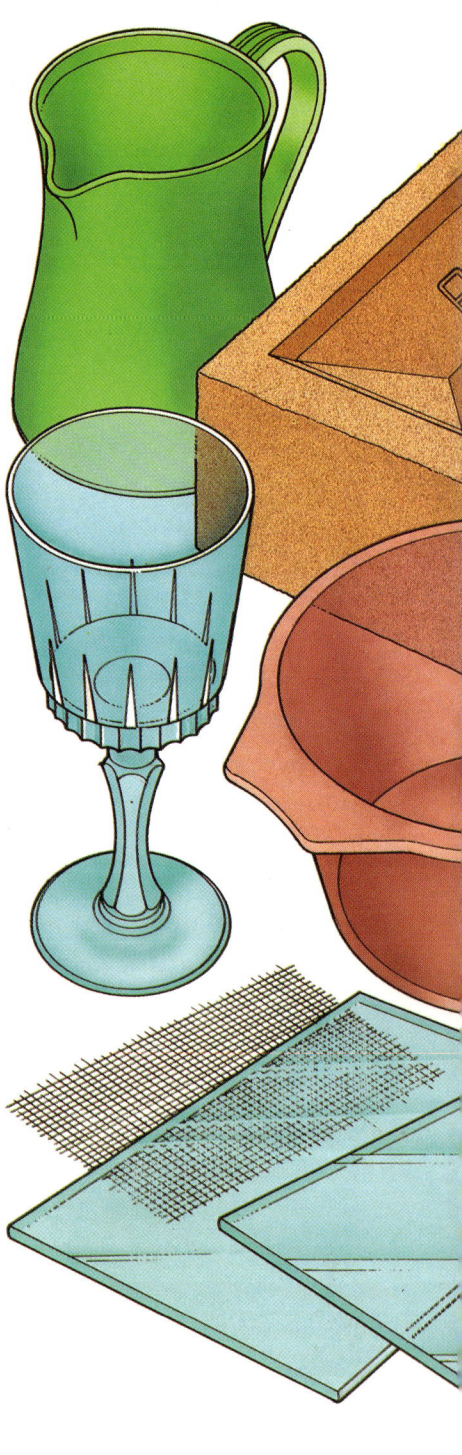

We think of glass as hard and not very bendable. Yet it can be drawn out into thin threads that are so flexible they can be woven into cloth and can be used effectively as electrical insulating tape.

Fibers of glass that has had all its impurities removed are very clear and will pass light along inside the fiber as if it were being piped through a tube. The light loses none of its brightness, even over distances of many miles. These optical fibers have many uses. In surgery, they enable doctors to look inside people without performing operations. In telecommunications, many separate messages can be sent along a single fiber at the same time and at the speed of light.

Ceramics include pottery and bricks of many kinds. Some modern ceramics are capable of withstanding much higher temperatures than metals. They are also very hard and brittle. Ceramic materials are used to make the heat shields of spacecraft that withstand searing temperatures on re-entering the atmosphere.

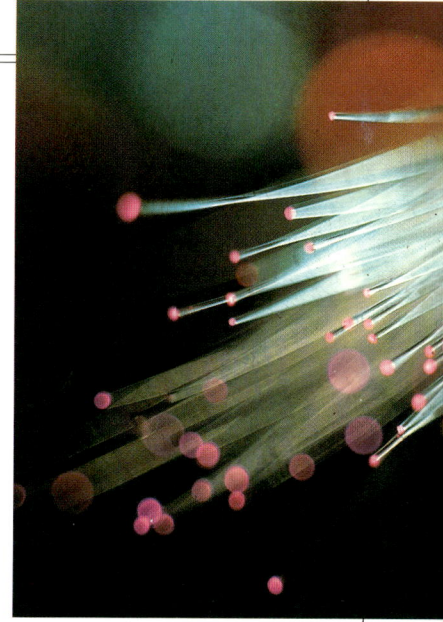

△ A multi-core cable containing optical fibers is capable of carrying thousands of messages at once. The cable is just like an electrical cable containing metal wires, but it can carry many more messages.

◁ When glass is drawn into very thin fibers and the fibers are all compacted together, often with a plastic, they form a very light and very tough insulating material called fiberglass, which has many applications. The hull of this boat is made entirely of fiberglass.

Water

Water is a liquid with many special properties. For example it can contain about ten times as much heat as can any metal. It is therefore very useful for storing heat produced by solar panels on houses, and also for keeping things cold – as in ice packs for picnics.

In industry water is used to cool a wide range of objects from machine-cutting tools to electric motors, from milk processors to hot rolled metal. Water is boiled to produce the high pressure steam required to drive steam turbines in power stations.

Water is so important in the home for drinking, cleaning and waste disposal that almost every house in the westernized nations now has a pure and permanent water supply delivered along an underground pipe.

△ Water alone is not an ideal cleaning liquid because it will not mix with grease, which makes dirt stick to things. Substances like soap and detergents are added to water to absorb grease and remove dirt. The molecules of soap or a detergent surround particles of grease and dirt and carry them into the water.

◁ High-pressure steam is used to drive the large turbines which in turn drive electric generators. The steam emerges at low pressure but is still very hot, and will accumulate in huge clouds unless it is condensed. One way of doing this is to mix it with a fine spray of water which must then be cooled in towers, like the ones shown here.

◁ A car radiator is filled with water, which cools the engine when it gets hot. The water passes through the engine and carries away heat. The hot water then flows through the radiator to lose this heat, and returns to the engine.

▽ Hydroponics is a way of growing plants without soil. Fed with water, mixed only with the nutrients essential to growth, grass over 15 cm (6 in) high can be grown in 8 days. Here we see tomatoes being grown using hydroponics. This emphasizes the importance of water in the life system that exists on our planet.

The vast amounts of water on the Earth distribute the Sun's energy. The heat from the Sun warms the ocean in the tropics, and warm currents then bring this heat to cooler regions. The Sun's heat also evaporates water and the water vapor forms clouds which contain water droplets. Clouds drop their water as rain, hail and snow on to high ground, from where it runs back in rivers to the sea. Dams are built in the rivers to store the water, which can then be allowed to flow through water turbines that drive electric generators.

Water is essential to all plant growth as well as to animal life. Our bodies — like those of most other mammals, birds, reptiles and fishes — are made up largely of water. The deadly effects of a lack of liquid water are to be seen in deserts, and in polar regions, where all the water is locked up in ice. Fishes, of course, can live only when they are totally immersed in water.

Gases

Air is a mixture of gases — mainly nitrogen, oxygen, argon and carbon dioxide. People and animals need oxygen to stay alive; plants need nitrogen, so a balance is struck which keeps the proportions of these gases in air about the same at all times.

The oxygen in air allows things to burn. We make use of it whenever we burn fuel, as in furnaces, or car and aircraft engines, and when we destroy waste by burning. Often the fuel is another gas such as that made from coal, natural gas from under the ground, or gasoline from crude oil. In a gasoline engine it is the expanding air and not the burning fuel that pushes the pistons to make power. The fuel is only the means of making the air hot.

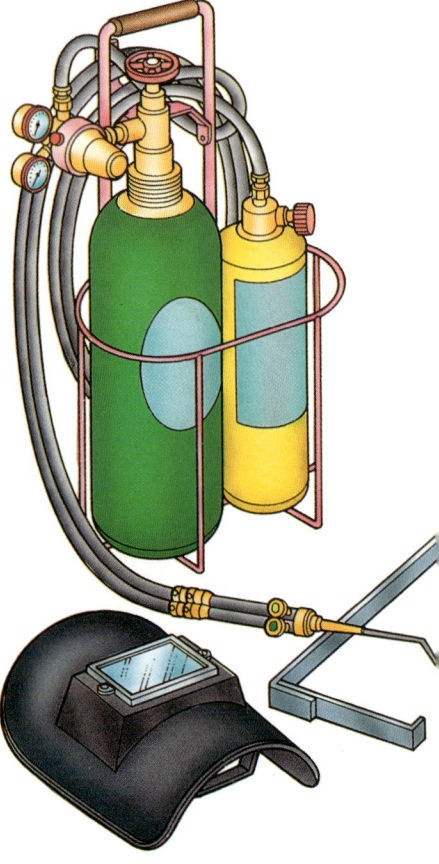

△ In welding, metal surfaces to be joined are melted and more raw metal is also added. This requires great heat, which can be supplied by burning a mixture of oxygen and acetylene gases in a jet.

◁ Neon lights are made of glass tubes containing neon or other gases that glow brightly when an electric current is passed through them.

Air, like water, is used for cooling engines and electric motors. A flow of air can also drive jackhammer drills and dental drills. Vacuum cleaners suck in air to lift dirt from floors and carpets.

Car tires are filled with compressed air to give a smooth ride. Air-filled mattresses enable us to sleep comfortably while camping. Large amounts of air and other gases can be stored by compressing the gas and piping it into cylinders. Compressed air cylinders enable us to go deep-sea diving and to stay under water for long periods. Some fire extinguishers contain compressed carbon dioxide gas.

Chlorine is a gas that is bubbled through water to kill harmful bacteria and purify the water. Helium is a very light gas used to fill balloons and airships.

Liquefied gases, such as liquid nitrogen and liquid oxygen, and the solidified gas carbon dioxide are very cold. They are used for achieving very low temperatures in scientific research.

△ Pottery is made by heating clay to a high temperatures inside an oven called a kiln. The heat is made by burning gas.

▽ During operations, patients may need to be unconscious for long periods. This can be achieved by having the patient breathe a mixture of oxygen and a pain-killing gas such as nitrous oxide.

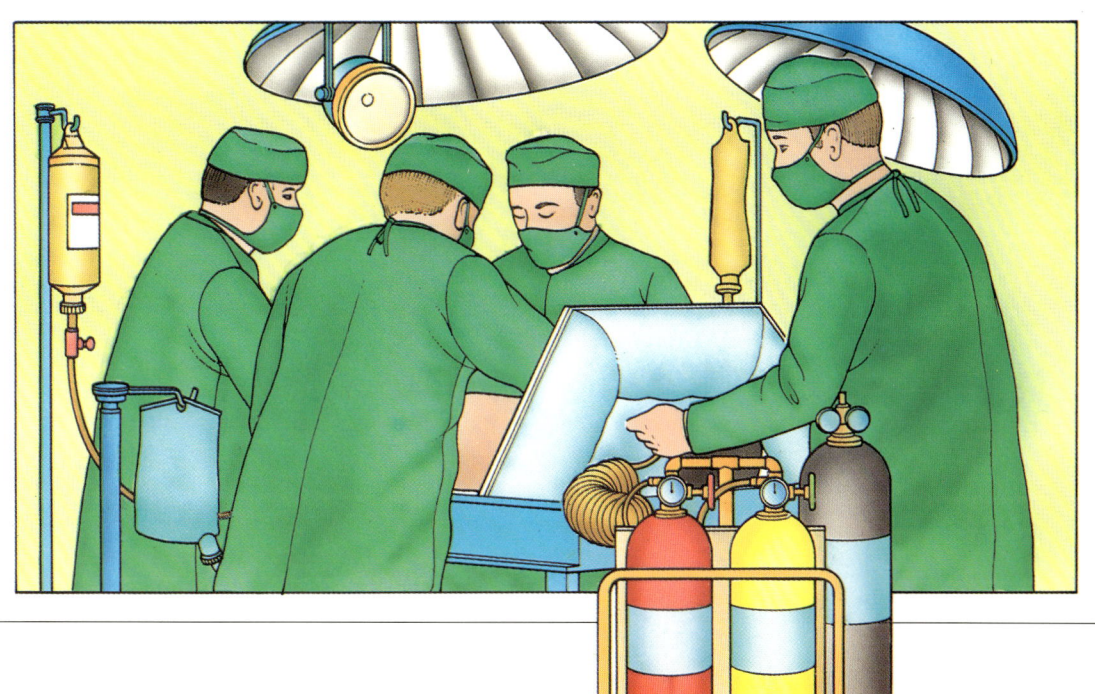

Materials that grow

Many of the materials that we use come from minerals in the ground or from air or from water. However, we also make use of living things, which we grow or raise. The most important of these are the crops and domestic animals that supply us with our food and drink.

Not all materials that we grow are used for food. Some plants are grown to provide oils for products such as soap and paint. Other important plant materials include rubber and plant fibers, such as cotton for making cloth. Animals also provide useful materials. Sheep give us wool, and the skins of many animals are treated with chemicals to make leather.

△ Many of our clothes are made of materials such as cotton, wool and leather that come from plants or animals.

◁ Small bottles are filled with drugs made by biotechnology in a chamber sealed to prevent contamination by germs.

▷ A farmer puts chemical fertilizer into a spreading machine. Such fertilizers have revolutionized farming, resulting in higher yields and allowing farmers to grow healthy crops on naturally poor soil.

Plants and animals are raised in fields and farms and harvested for use. However, several useful things are grown in laboratories or factories and used to make foodstuffs and drugs. Growing living things in artificial environments is known as biotechnology.

Bacteria may be grown in laboratories and are used in the production of many foods and drinks. This process is called fermentation. Cheese and yogurt come from fermented milk, for example, and alcoholic drinks such as beer and wine come from plants such as cereals and grapes. Bacteria grown in the milk or yeast in the juices from the plants change them into these foods and drinks. Alcohol made by fermentation of plants such as sugarcane can be used as a fuel in car engines.

▽ Many drugs are obtained from living things. Digitalis is found in the common foxglove. The drug industry is a development from the ancient practice of using herbal medicines to cure people.

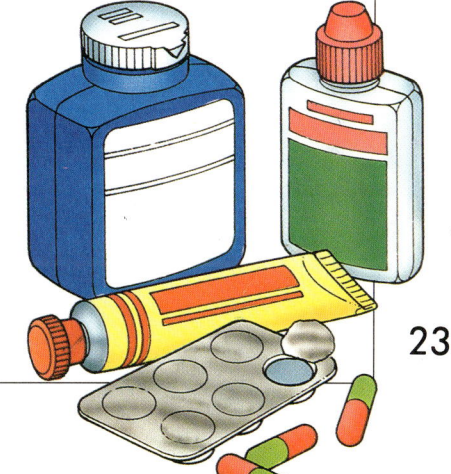

Fossil materials

Fossils are more than the remains of individual prehistoric animals and plants. The tiniest creatures living in the sea were massed together in huge quantities when they died, and were then covered in mud. Plants and trees were buried in the ground millions of years ago. These remains were then subjected to huge pressures and high temperatures as the result of volcanic action and the movement of the Earth's crust, which changed them dramatically. Coal formed from the wood in trees. Oil formed from the minute animals and plants. Natural gas forms as a product of the decay of living matter.

Crude oil, as naturally obtained from the ground, is thick and black and contains many valuable chemicals. To make use of the basic material, the oil itself, it must first be refined.

Here we see two fossil materials in use: oil products flowing through the pipes of an oil refinery (above), and coal piled in heaps for burning in a power station (below). Both oil and coal, as well as being valuable fuels, are important starting materials for making a wide range of chemical products.

Oil can be refined to make the clear liquid gasoline, the most popular fuel for cars. Diesel engines use a heavier fuel, also obtained from oil.

Natural gas from under the sea bed is piped ashore and distributed to homes as an alternative to coal gas.

Coal is one of the most useful fossil materials. The gas made from coal, like oil, contains many valuable chemicals which can be washed out of it — the gas is "scrubbed" in water — before the gas itself is supplied as a fuel to factories and to industry. The washing water then contains materials which are used to provide cosmetics, dyes, paints, plastics, fertilizers and medicines. Coal is also converted to a liquid called coal tar that is another important source of chemicals.

When we use fossil materials we are working in the best traditions of nature, which recycles (reuses) waste materials produced by the natural resources of the Earth. In nature, for example, trees shed their leaves, which decay in the soil to provide fertilizer for the next year's growth.

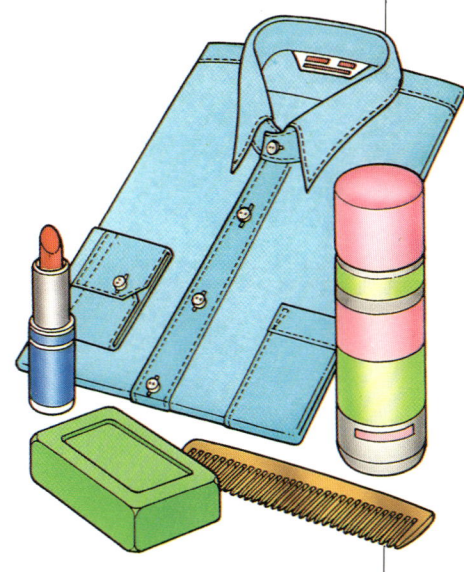

It is difficult to imagine that many household objects like those shown above come from coal or oil. It is also a strange fact that when we paint our homes, we are using a material that formed from living creatures that inhabited the Earth millions of years ago.

Chemicals

▽ Many parts of a modern car, revealed in this special model, are made of plastic and other synthetic materials produced by chemical engineering. The paint, lubricating oil, and fuel of the car itself are the result of refinements in materials — all of which are made possible by the chemical industry.

△ The special aircraft *Voyager* flew around the world non-stop. It was built of light plastic materials that were strong enough for the whole hollow structure to be filled with fuel, right to the wing tips, at the beginning of the flight.

Chemistry has enabled us to produce new materials, extending traditional ones by thousands of times. Chemistry therefore effects all our lives every day. Even cooking is a form of chemistry.

We use chemicals in medicines and in making the most advanced drugs. We also use them in making fearsome weapons of war such as shells, mines, bombs, and poison gas.

Among the less dramatic uses of chemicals is the production of chemically made colors: dyes for fabrics, paints for preserving and brightening up our houses, machinery, cars and toys. There are colors for artists' paints, colors for cosmetics and colors for foodstuffs. These artificial colors are much brighter than natural dyes.

▽ Many of the products we eat and drink contain artificial flavors and colors made by chemistry. Carbonated drinks also contain the gas carbon dioxide.

Food growing has been greatly improved by the use of chemical fertilizers, insect pest sprays and weed killers. These and other chemicals are produced on such a large scale that the profession of chemist has extended to that of chemical engineer.

The whole plastics industry is founded on the application of chemistry. Plastics include the synthetic fibers of which so many of our clothes are now made. The modern kitchen is full of articles made from plastic. It is possible to buy a picnic outfit made entirely of plastic materials.

Chemical engineers can use chemistry to "design" new materials for special uses. The *Voyager* aircraft that flew nonstop around the world in 1986 was built of special plastic that is very light, enabling it to carry lots of fuel. New synthetic fibers can give us clothes that are fireproof and even bullet-proof!

▽ A combination of lightness and strength is also needed in a modern ocean-racing yacht. The hull has a special plastic skin, and the ropes and sails are made of synthetic fibers.

Modern materials

Materials science is a relatively new subject which includes ceramics, plastics and the study of metals. More and more elements are being exploited in every generation to provide new materials. Germanium is common in ashes from spent fires. When purified to a degree once thought impossible, it produced the transistors that made the miniature portable radio possible. Silicon is the most common solid element on Earth and the basic substance in dust on the streets. Again, when highly purified, it yielded the silicon chip at the heart of the computer.

As well as exploiting little used elements, modern materials are also made by using existing ones in different ways. Reinforced materials are special materials of great strength. Many are made of plastics that contain layers of carbon fibers. The fibers are very strong while the plastic is light, giving a material suitable for aircraft wings, helicopter rotor blades, skis and tennis rackets.

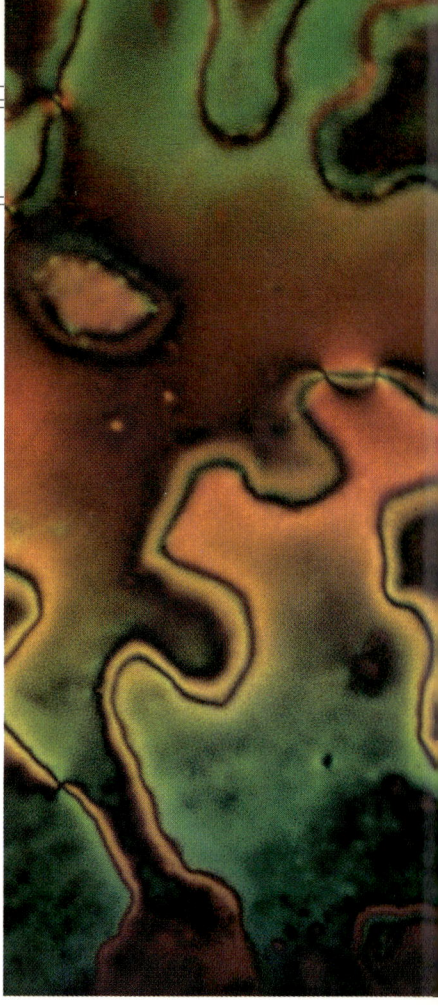

△ A view of a liquid crystal magnified 100 times. Rearrangement of the molecules in the crystal affects the way in which light passes through the crystal, producing various colors or causing a display to become dark or transparent.

◁ The wings of this advanced aircraft are made of plastic reinforced with carbon fibers. This material gives maximum strength with minimum weight. It can be applied in layers so that complex shapes can be easily built up.

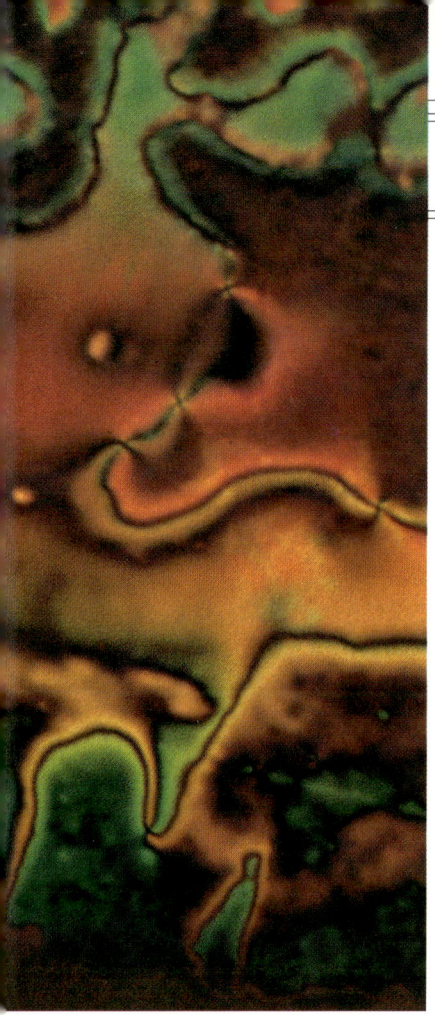

Advances in materials have made seemingly impossible ideas become facts. For example, developments in ceramics for gas turbine blades provided the propulsion needed for supersonic flight. New materials were also needed for spaceflight — not only to get spacecraft into space but also to get them back. Because re-entry into the Earth's atmosphere produces searing heat, special materials were invented to shield spacecraft. A thin piece of the heat-resistant ceramic used on the Space Shuttle can be red-hot on one side, yet cool enough to touch on the other!

A number of new materials have been produced by developing new kinds of matter. The liquid crystals in digital displays on watches and calculators are an example. Although liquid, their molecules line up in rows like those of solid crystals. The molecules turn when subjected to heat or an electric current, causing the material to change color.

▷ Many new materials have been developed for use in space. Here technicians inspect heat-resistant tiles on the space shuttle *Columbia*. Later versions of the shuttle will include further developments in tiles to withstand even higher temperatures.

Glossary

Alloy
A material made of two or more elements that are metals, and possibly some elements that are not metals. The metals are melted and mixed together, and then cooled to form an alloy. The strength and other properties of most alloys are very different from the metals of which they are composed.

Ceramic
A material that is made from various kinds of clay by heating the clay to a high temperature in a kiln. The particles in the clay bond together and it cools to form a material that is hard but brittle, such as pottery, porcelain, brick and tile. Ceramics are very heat-resistant and good electrical insulators. Cermets are tough ceramics that contain metal.

Chemical
A chemical or chemical compound is a substance that is composed of a particular combination of elements. Salt is a chemical made of the elements sodium and chlorine. Sugar is another chemical containing carbon, hydrogen and oxygen. The chemical compound is unlike the elements of which it is composed. Chemicals can be made from the pure elements, or by causing other compounds to react with each other so that they exchange elements. Chemicals can also be extracted from natural materials such as oil and coal.

Concrete
A building material made by mixing together sand, cement, gravel and water. It dries to form a hard material in which the three solid substances are bound strongly together. The cement, which has the smallest particles, fills the spaces between the particles of sand, which in turn fill the spaces between the pieces of gravel.

Conductor
A material through which an electric current can flow. The element carbon and several metals are good conductors. An electric current consists of a flow of tiny particles called electrons. Conductors are made up of atoms (*see* Molecule) in which electrons are held loosely. When a current flows, the electrons pass easily from one atom to the next.

Electromagnet
A kind of magnet that becomes strongly magnetic when an electric current flows through it. The electromagnet basically contains a core of iron surrounded by a coil of wire. When a current flows through the coil, it creates a powerful magnetic field. When the current stops, the core loses most of its magnetism.

Element
A material that contains only one kind of atom in its molecules. Just under 100 elements occur in nature, and a few more artificial elements can be made, such as plutonium. Common natural elements include oxygen, carbon, iron, gold, silver, copper, chlorine, aluminum, hydrogen and nitrogen. Most elements occur in chemical compounds.

Fiber
A thin strand of material. Short natural fibers are obtained from plants such as cotton, and from the hair of animals such as sheep. These fibers are spun together to make thread or yarn, which is then woven or knitted to produce fabrics. Synthetic fibers are fibers of plastics that are made from chemicals. They are spun and woven to produce synthetic fabrics, such as nylon and polyester.

Fuel
A material that is burned to provide heat. Natural fuels include materials found in or on the ground such as wood, coal and natural gas. Many machines need artificial fuels made from oil, such as gasoline for cars and diesel fuel for trucks and trains.

Insulator
A material that resists the flow of heat or electricity. Heat insulators include wool and fluffy materials, which trap air between the fibers. These are used to line coats and for blankets. They prevent us from losing body heat, keeping us warm. Electrical insulators are used to make electrical switches and plugs, and to coat cables and wires. Plastics and ceramics are also good electrical insulators.

Microchip
A tiny electrical device used in computers, calculators and many other electronic machines. A microchip contains many thousands of miniature transistors and other electrical parts made of semiconductors. Electric code signals move in and out of these parts to make the microchip work. It can store the signals or make calculations with them.

Molecules
The tiny particles of which materials are made. A molecule is a group of even tinier particles called atoms connected together. Each atom belongs to a certain element. Most molecules have atoms of two or more elements. A grain of sugar contains about a billion billion molecules. Each one is made of 45 atoms of carbon, hydrogen and oxygen.

Oil
Crude oil is a thick dark liquid material found in the ground or under the sea. Fuel such as gasoline and useful chemicals can be made from crude oil. Lubricating oil is another kind of oil that is used in engines; it is also made from crude oil. Some fuels are also called oil and so are liquids extracted from plants (such as olive oil).

Ore
A material found in the ground from which metal can be extracted. An ore contains a mineral, a chemical compound containing the metal and one or more other elements, often oxygen. Bauxite is an ore of aluminum, for example. Intense heat is usually needed to extract metals from ores.

Plastics
Materials that are made from chemicals and have big molecules. The molecules consist of long chains of carbon atoms joined together. In many plastics, the chains can bend, making these materials flexible. In others, the chains are linked together so that the plastics are hard and tough.

Semiconductor
A material that is used to make microchips and transistors. It usually consists of the element silicon to which small amounts of other elements are added. These elements make the semiconductor pass varying amounts of electric current so that it is able to handle electric signals.

Transistor
An electrical device made of several pieces of semiconductor. An electric signal goes to the transistor, which causes the semiconductors to pass or block an electric current. In this way, the transistor can act as a switch or it can make the signal stronger.

Index

air 4, 9, 20–1
aircraft 7, 14, 20, 27, 28
alcohol 23
alloy 11, 30
aluminum 11, 13
animal products 5, 22
argon 20

bacteria 23
biotechnology 23
brass 11, 13
brick 6, 16, 17
bronze 11

cadmium 13
car 7, 10, 18, 20, 21, 25, 26
carbon 13
carbon dioxide 20, 21
carbon fiber 28
cardboard 6, 15
cement 12
ceramics 6, 7, 16–17, 29, 30
chemical engineering 5, 27
chemicals 6, 24, 25, 26–7, 30
chromium 11, 13
clay 6, 21
clothing 6, 22, 27
coal 5, 20, 24, 25
cobalt 13
concrete 11, 30
conductor 7, 12, 30
copper 11, 12, 13
cotton 5, 6, 22
crystal 8, 9, 28, 29

diesel 25
drugs 23, 26
dyes 25, 26

electric generator 13, 19
electric motor 7, 12, 18, 21
electromagnet 13, 30
element 30

fertilizer 22, 25
fiberglass 6, 17
fibers 5, 6, 17, 27, 30
food 22, 23, 26, 27
fossil fuels 24–5, 30

gas (fuel) 5, 20, 21, 24, 25
gases 8–9, 20, 21, 26
gasoline 6, 20, 25
germanium 28
glass 6, 8, 9, 16–17
Giotto spaceprobe 7
gold 10, 12

helium 21
hydroponics 19

ice 8, 9
insulation 6, 17
insulator 7, 12, 30
iron 11, 12, 13

jet engine 17

lead 11, 13
leather 22
liquid crystals 28, 29
liquids 8–9, 11, 13, 16

magnesium 11
magnetism 7, 12, 13
matter, states of, 8–9
mercury 13
metals 6, 7, 8, 9, 10–11, 12–13, 14, 16, 18, 28
molds 8, 13, 16
molecules 8, 18, 28, 29, 30

natural gas 24, 25
neon 20
newsprint 15
nickel 11, 13
nitrogen 20, 21
nylon 6

oil 5, 6, 24, 25, 30
optical fibers 17
ores 5, 6, 11, 30
oxygen 12, 20, 21

paints 25, 26
paper 4, 6, 15
plant fibers 5, 22
plastics 6, 7, 12, 17, 25, 26, 27, 28, 30
platinum 10, 12
polyester 6
polystyrene 6
pottery 6, 17, 21

radio 7, 28
reinforced concrete 11

sand 5, 6
semiconductor 6, 30
silica 16
silicon 13, 28
silver 10
solar panels 18
solder 12, 13
solids 8–9
spacecraft 7, 17, 29
steel 10, 11, 12, 13
synthetics 6, 27

temperature, effect on materials, 4, 8, 9, 10, 11, 13, 16, 17, 21, 29
tin 9, 11, 13
transistor 28, 30
tungsten 13
turbine 16, 18, 19, 29

water 4, 8, 12, 18–19, 21
welding 11, 20
wood 4, 5, 6, 12, 14–15, 24
wool 5, 6, 22

zinc 11